19 Self-Correcting Activities

by
Dolores Freeberg

Good Apple

Editor: Susan Eddy
Designer: Lisa Ann Arcuri
Cover Illustrator: Terry Taylor

GOOD APPLE
An Imprint of Modern Curriculum
A Division of Simon & Schuster
299 Jefferson Road, P. O. Box 480
Parsippany, NJ 07054-0480

©1997 Good Apple. All rights reserved. Printed in the United States of America. The publisher hereby grants permission to reproduce these pages in whole or in part, for classroom use only.

ISBN 1-56417-684-3

1 2 3 4 5 6 7 8 9 VIC 01 00 99 98 97

CONTENTS

Introduction 4

Addition Skills

Puppy (addition 0–6) 6
Dinosaur (addition 0–6) 8
Teddy Bear (addition 0–6) 10
Whale (addition 5–10) 12
Elephant (addition 5–10) 14
Snowman (addition 5–10) 16
Turtle (addition 0–10) 18
Frog (addition 0–10) 20
Tugboat (addition 0–10) 22

Subtraction Skills

Jack-o'-lantern (subtraction 0–6) 24
Rocket (subtraction 0–6) 26
Turkey (subtraction 0–6) 28
Fish (subtraction 5–10) 30
Humpty Dumpty (subtraction 5–10) 32
Duck (subtraction 0–10) 34
Bee (subtraction 0–10) 36
School Bus (subtraction 0–10) 38

Mixed Skills

Train (addition and subtraction 0–10) 40
Flowers (addition and subtraction 0–10) 42

Answer Key 44

INTRODUCTION

What is Picture-Puzzle Math?

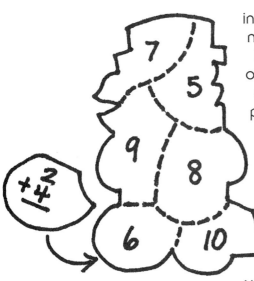

Picture-Puzzle Math is an innovative, sequential, self-correcting math puzzle activity book for children in kindergarten and first grade. By figuring out answers to the math problems found on each puzzle piece and placing these pieces on the corresponding correct answers, children form pictures that they may color.

The activities in *Picture-Puzzle Math* meet a variety of objectives. Children practice math concepts, reinforce spatial relationships, improve fine motor skills, and focus on single concepts as they work through the various puzzles.

What materials are required?

Each child will need the following items for each activity.
- puzzle background page
- puzzle piece page
- scissors
- white glue or glue stick
- pencil

The following student directions apply to each activity. You may wish to reproduce one puzzle and its pieces on overhead transparencies for demonstration purposes.

1. Carefully cut out each puzzle piece on the solid lines. (It may be easier for some children to separate all the pieces on the puzzle piece page, then trim carefully along the cutting lines.)

2. Figure out the answer to the math problem on each piece in your head. Don't write it down yet!

3. Without using any glue, check to see if the piece fits in the space on the background page that shows the answer you calculated. If it does, place it there without gluing. If it does not, rethink your answer and try again.

4. When you have figured out the correct answer, write it in pencil on the puzzle piece and place it on the background page. Do this until you have placed all the puzzle pieces. *Don't glue any pieces in place until you have found the right place for each piece.*

5. Apply white glue or glue stick to each piece and replace it in the spot you have chosen. Try to glue all edges tightly so your picture is flat.

6. When the glue has dried, color the completed puzzle.

Enrichment

You may wish to increase the challenge of these activities by whiting out the puzzle-piece outlines on the background pages. Children would then have to work a bit harder to ensure that all the pieces fit together correctly. Remind them to do no gluing until they are satisfied that each piece is correctly positioned within the picture outline.

Another enrichment possibility would be to white out problems and solutions on a puzzle of your choice and invite students to create their own puzzles. In this case, they would begin by placing the cut-out pieces within the dotted-line spaces. They would then write math problems on the puzzle piece and the corresponding answers in the spaces underneath—on the background sheet. Children may enjoy exchanging puzzles with partners to reassemble.

Name _____

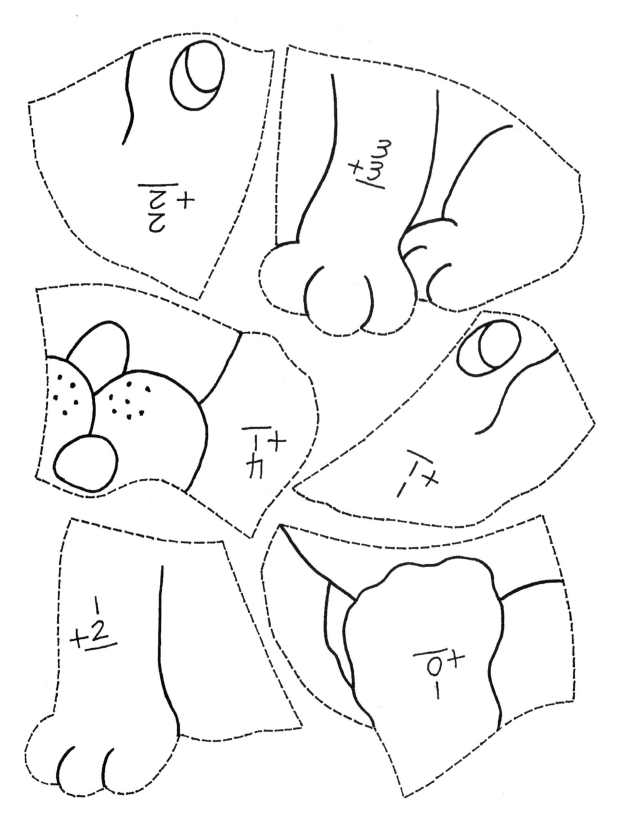

Addition 0–6

Name _____

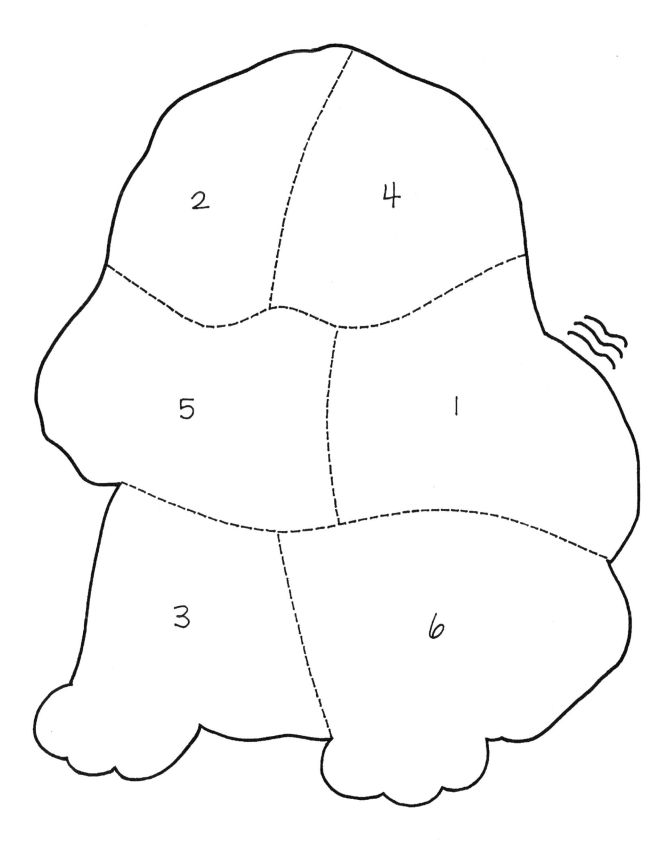

Addition 0–6

Name _____

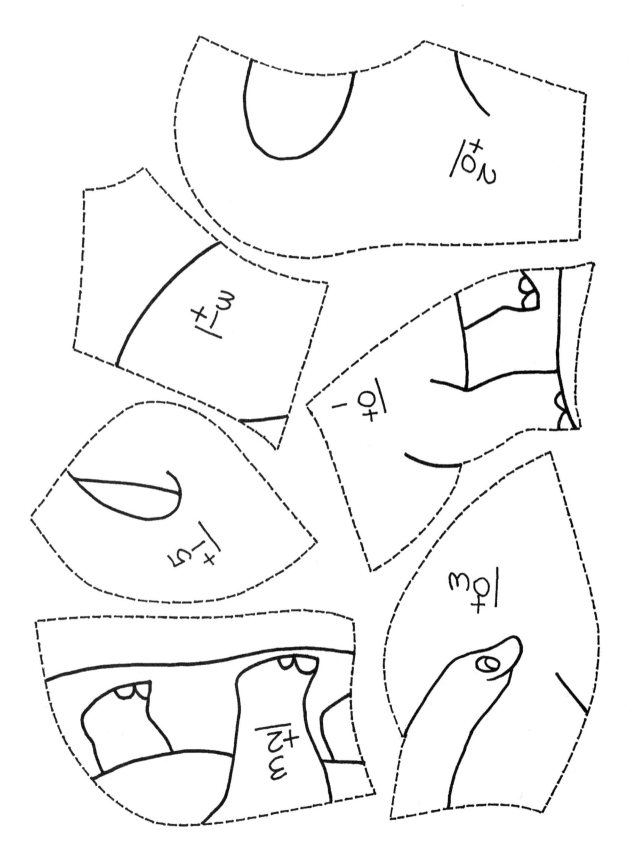

Addition 0–6

Name _____

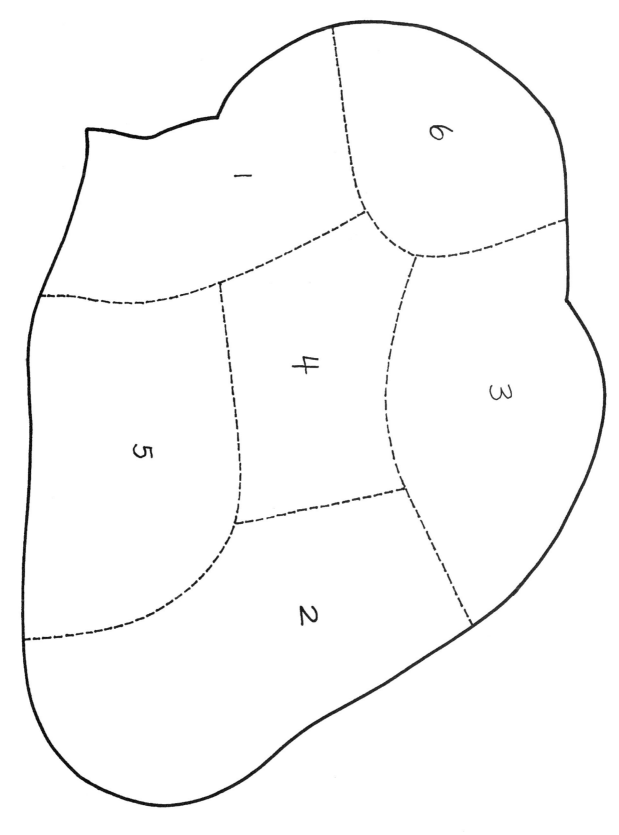

Addition 0–6

Name _____

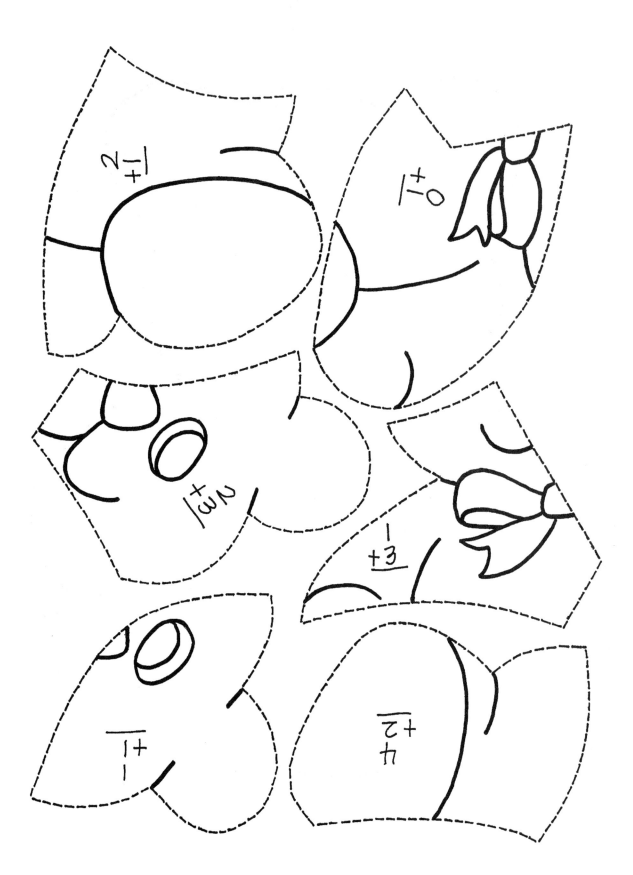

Addition 0–6

Name _____

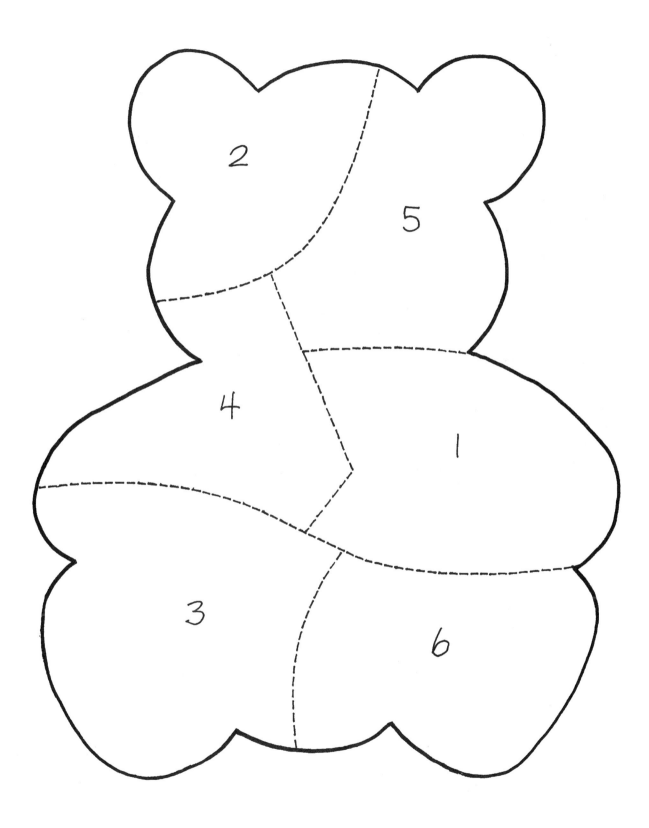

Addition 0–6

Name _____

Addition 5-10

Name _____

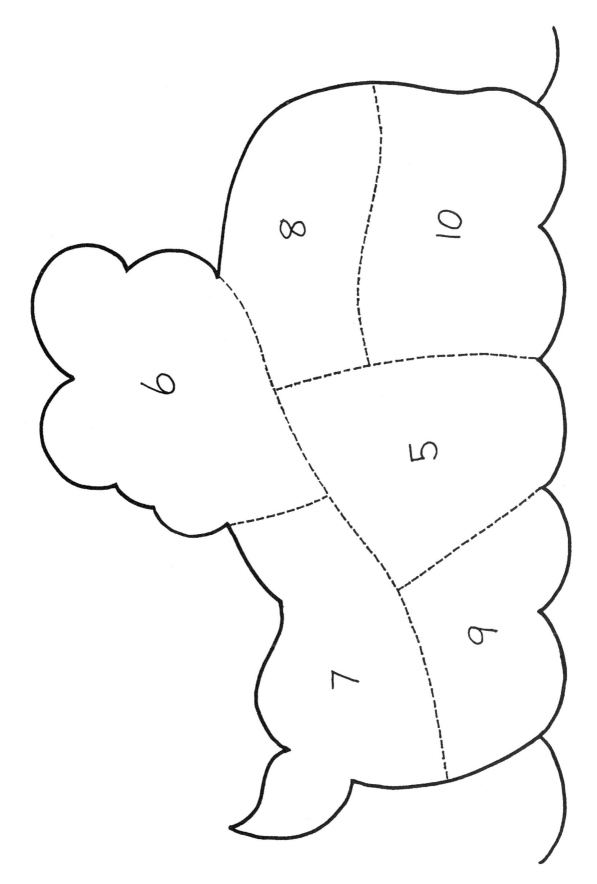

Addition 5-10

Name _____

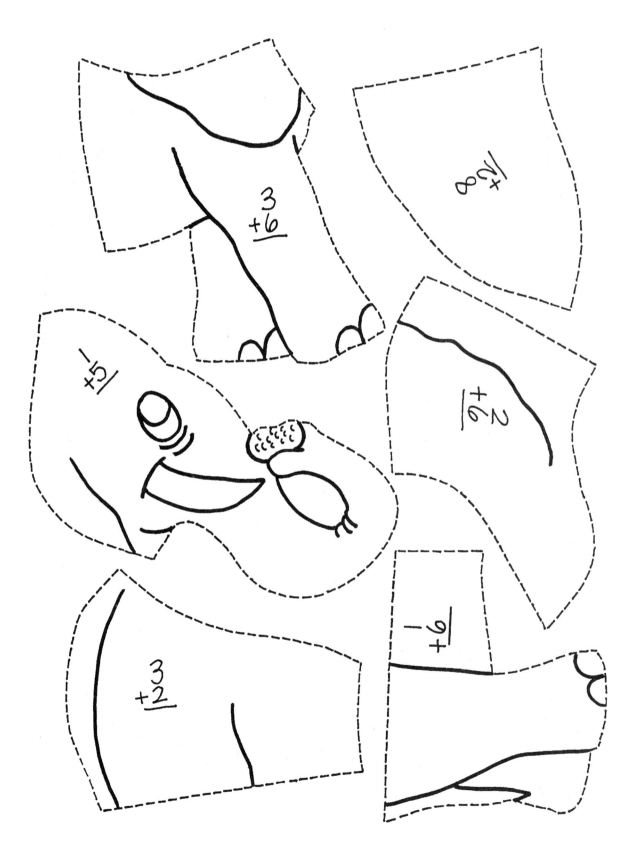

14 Addition 5-10

Name _____

Addition 5-10

Name _____

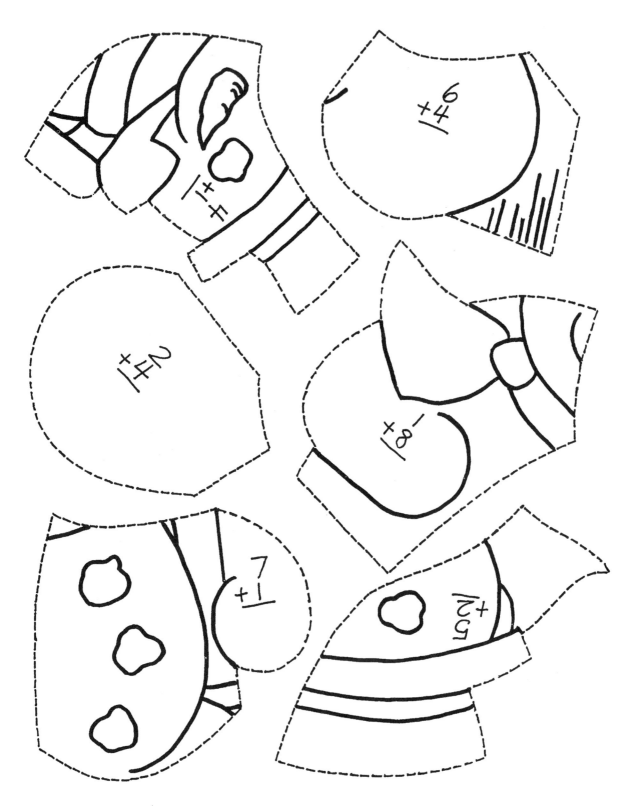

Addition 5-10

Name _____

Addition 5-10

Name _____

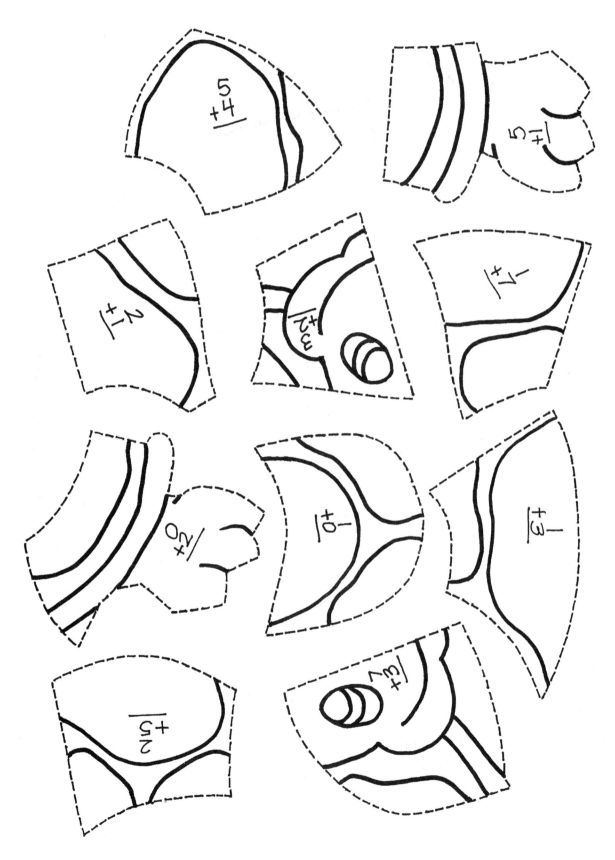

Addition 0-10

Name _____

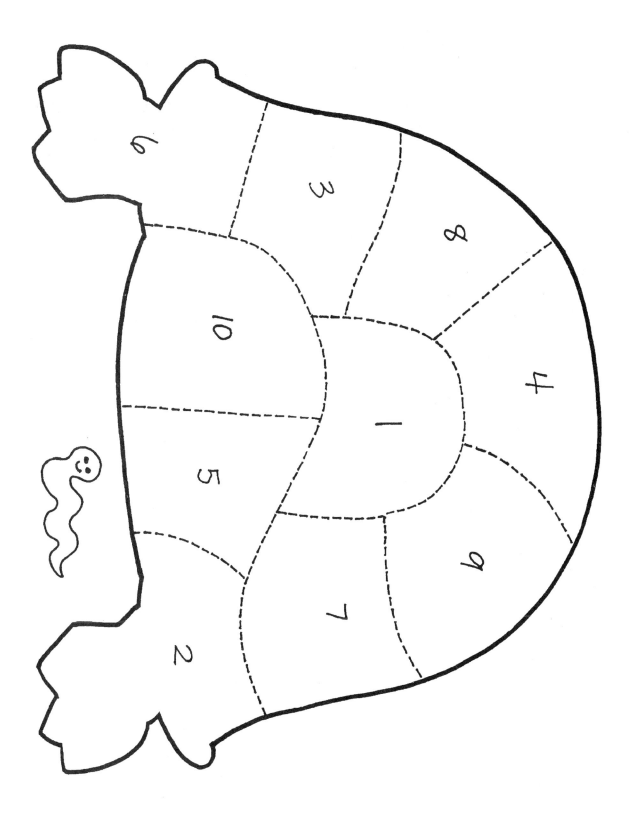

Addition 0-10

Name _____

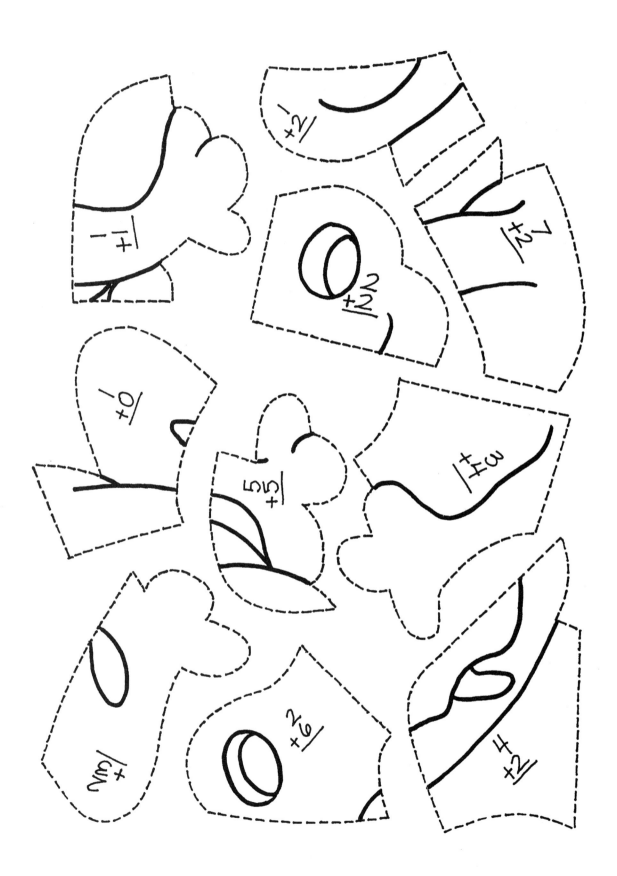

Addition 0–10

Name _____

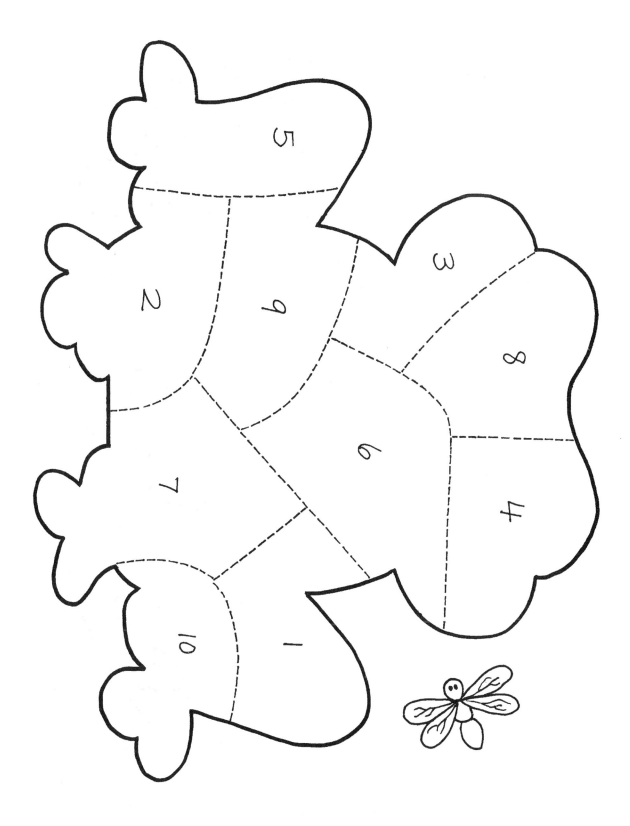

Addition 0–10

Name _____

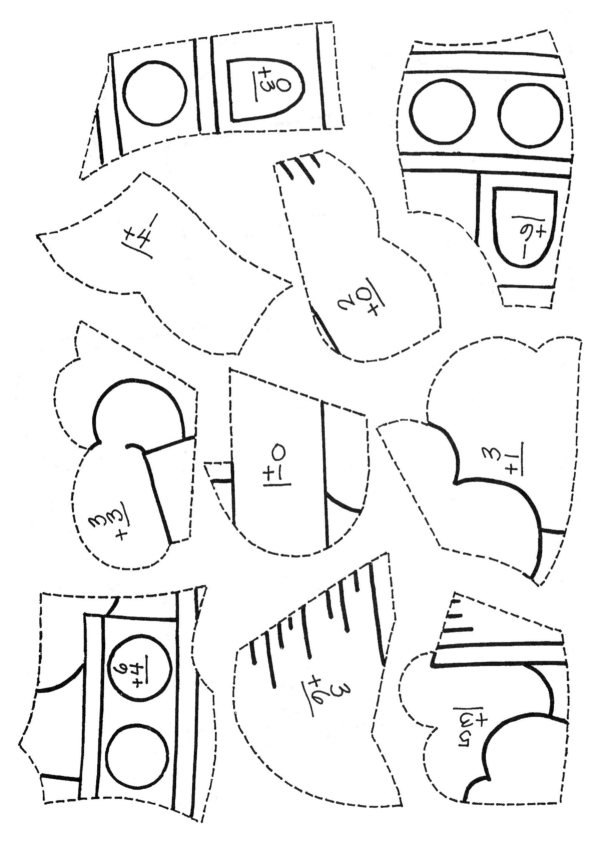

22 Addition 0–10

Name _____

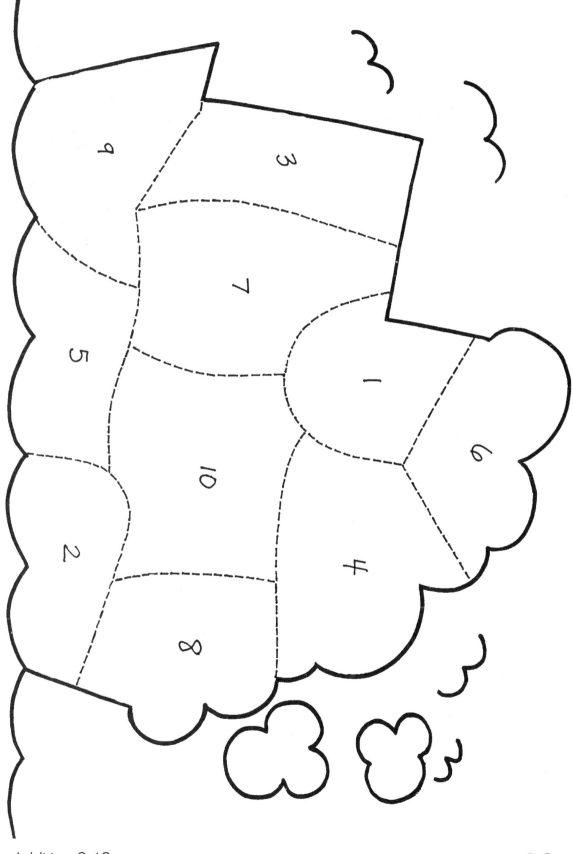

© 1997 Good Apple

Addition 0–10

23

Name _____

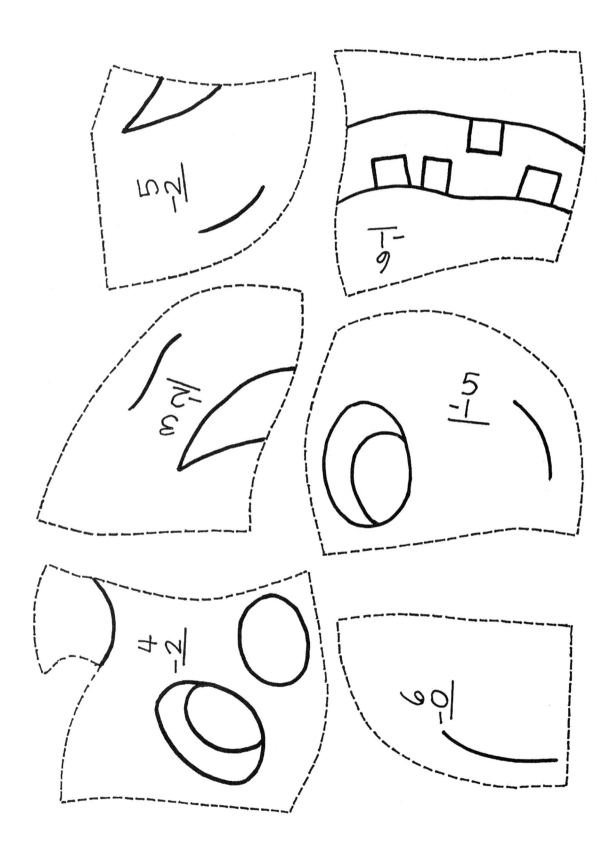

24   Subtraction 0-6

Name _____

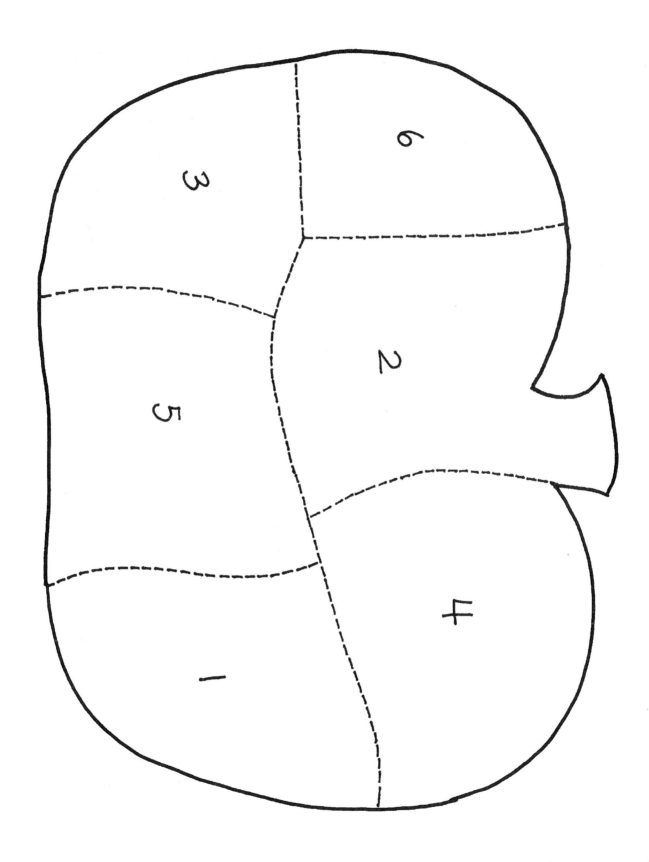

Subtraction 0-6

Name _____

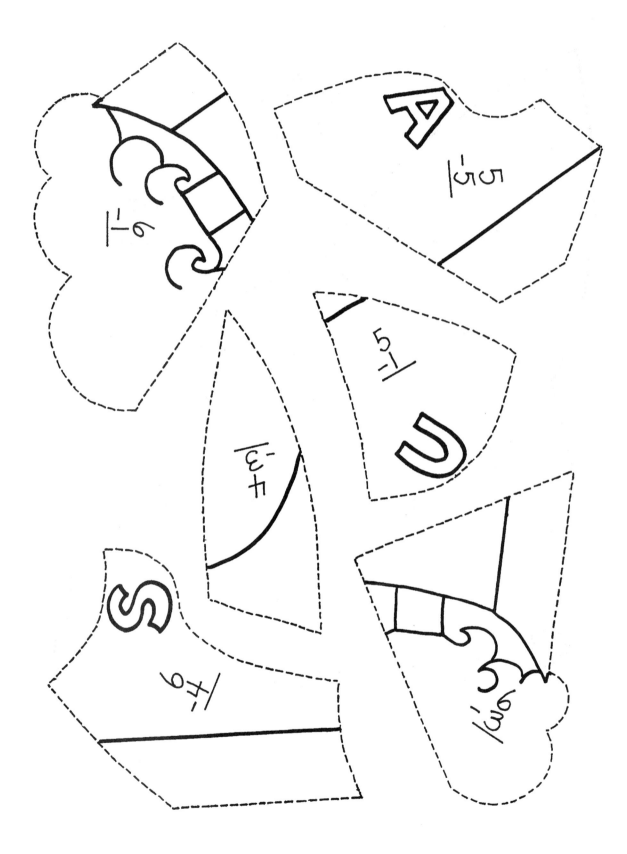

26 Subtraction 0-6

Name _____

Subtraction 0-6

Name _____

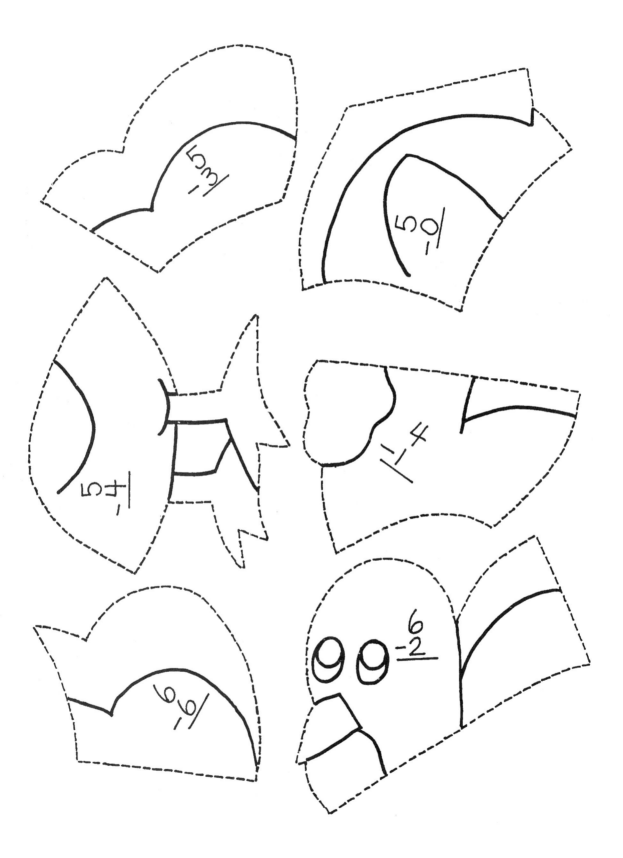

28 Subtraction 0–6

Name _____

Subtraction 0–6

Name _____

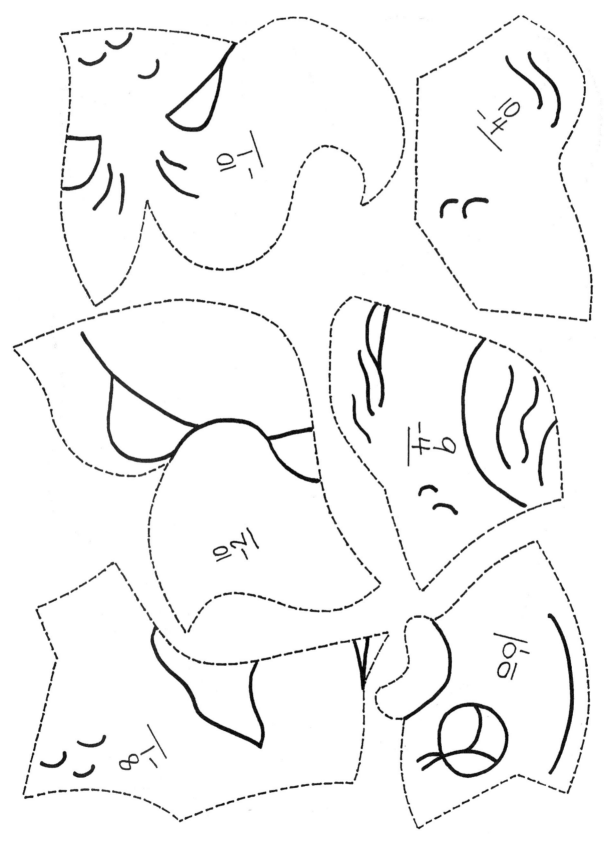

30　　　　　　　　　　　　　　　　　　　　　　　Subtraction 5-10

Name _____

Subtraction 5-10

Name _____

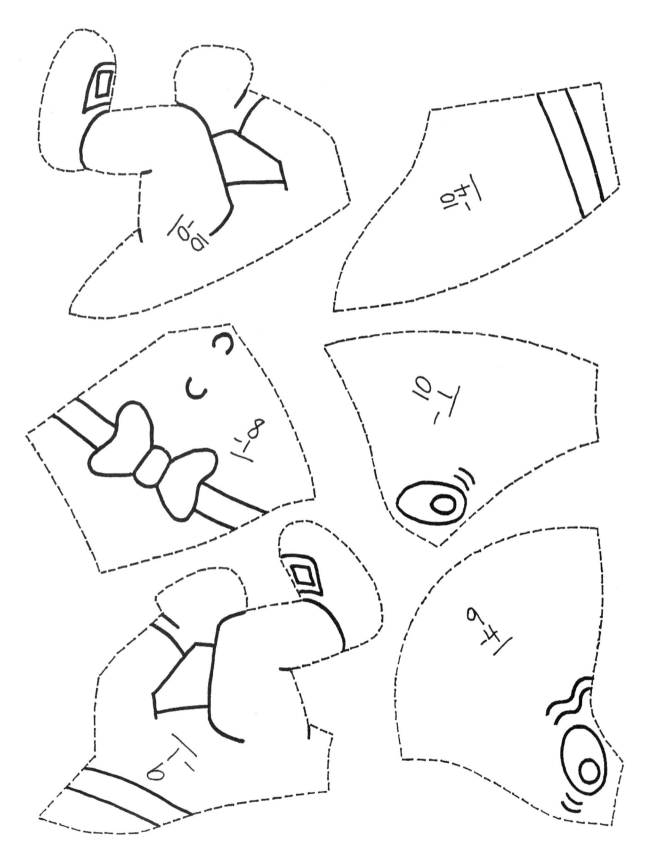

Subtraction 5-10

Name _____

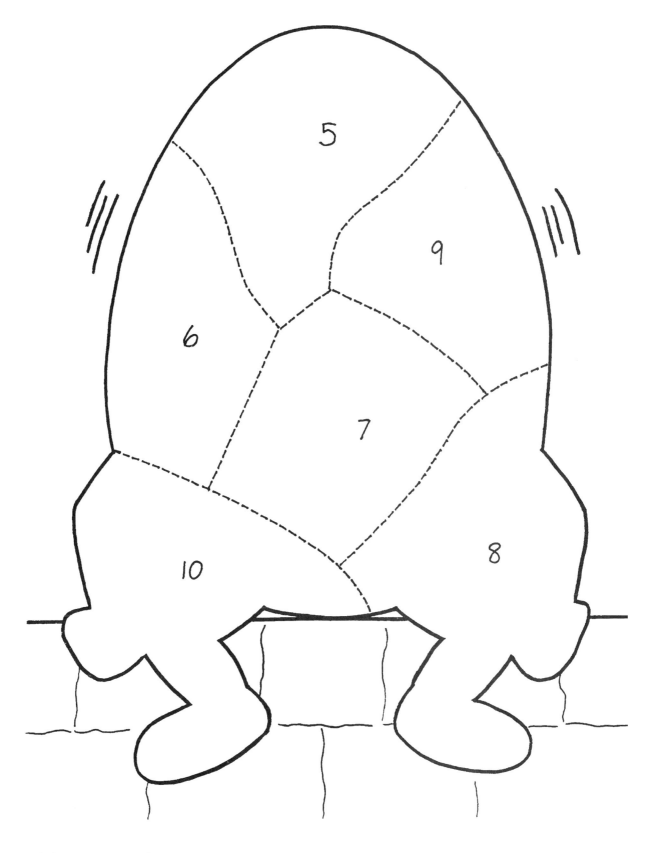

Subtraction 5-10

33

Name _____

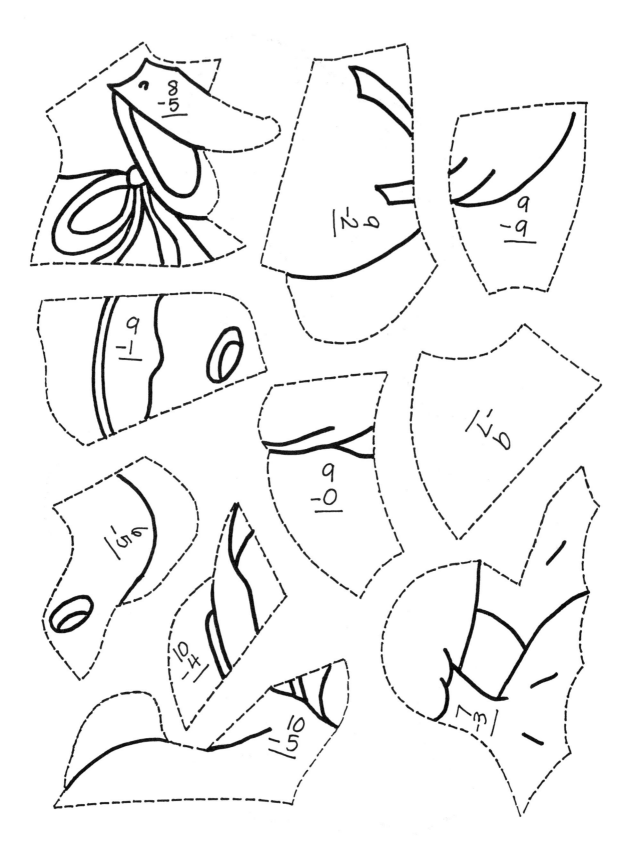

34 Subtraction 0-10

Name _____

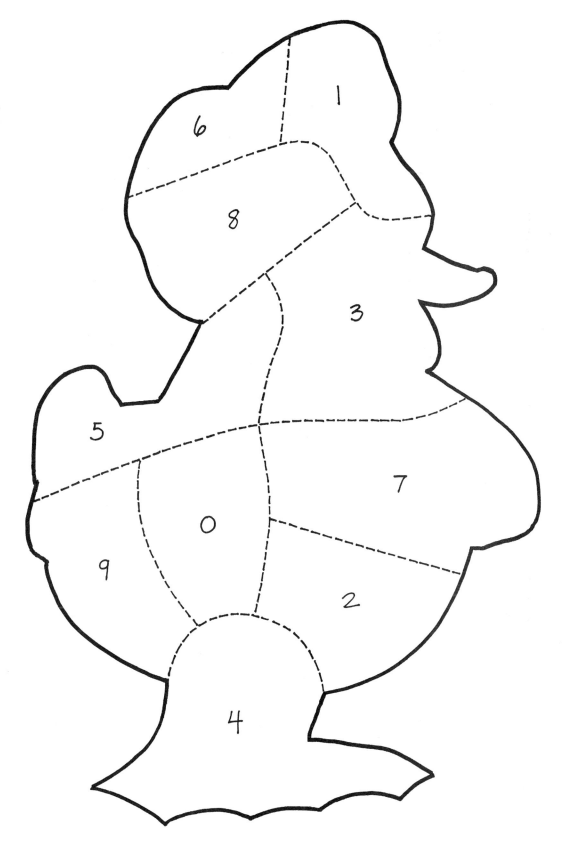

Subtraction 0-10

35

Name _____

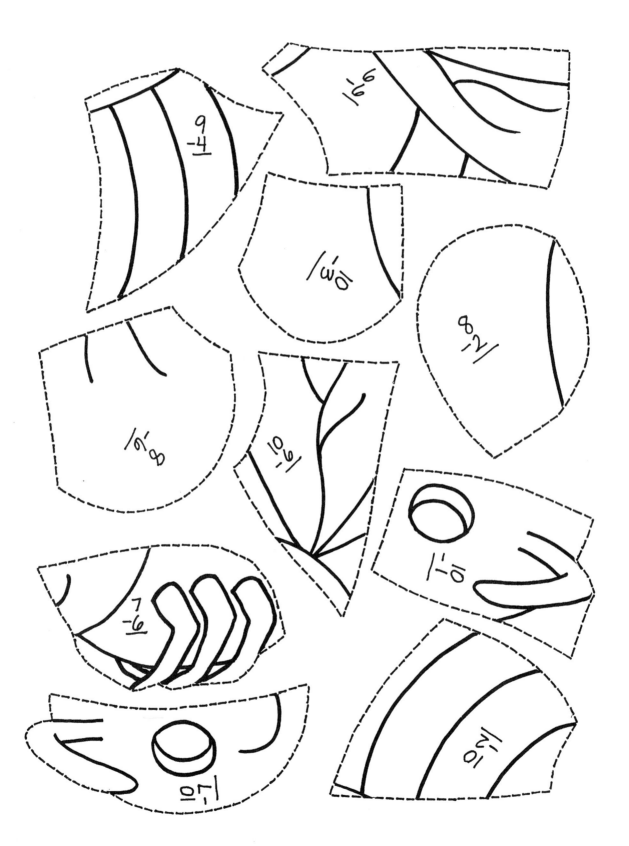

Subtraction 0-10

Name _____

Subtraction 0-10

Name _____

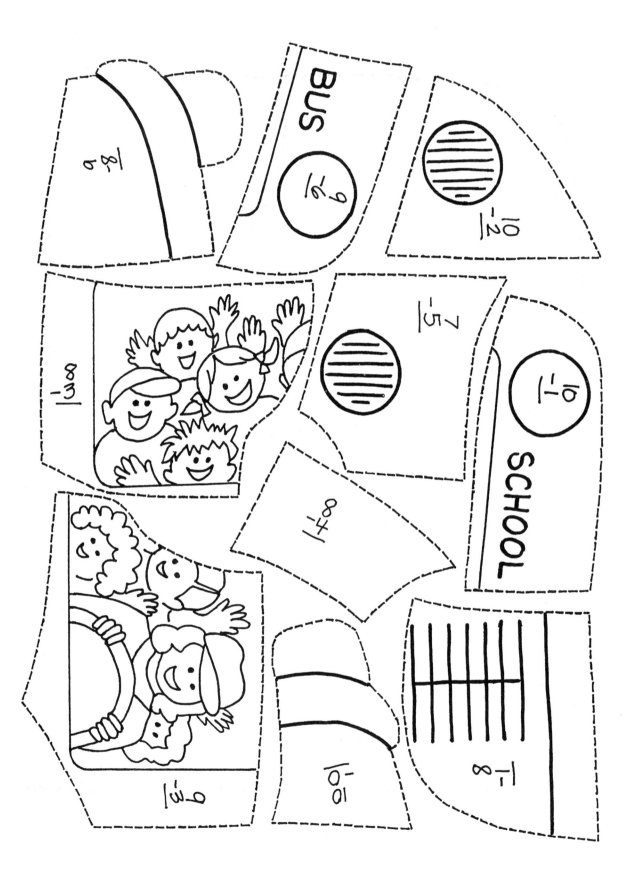

38 Subtraction 0-10

Name _____

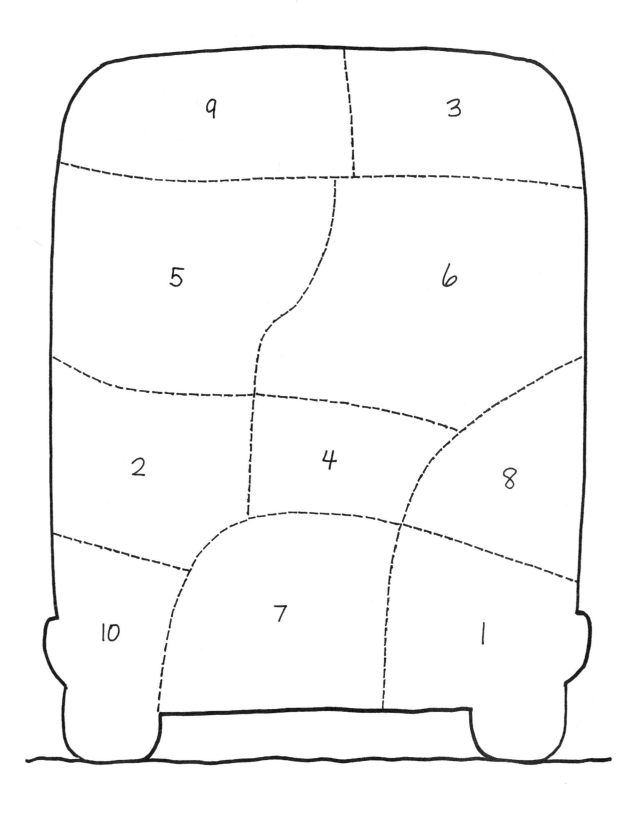

Subtraction 0-10

Name _____

40

Addition and Subtraction 0-10

Name _____

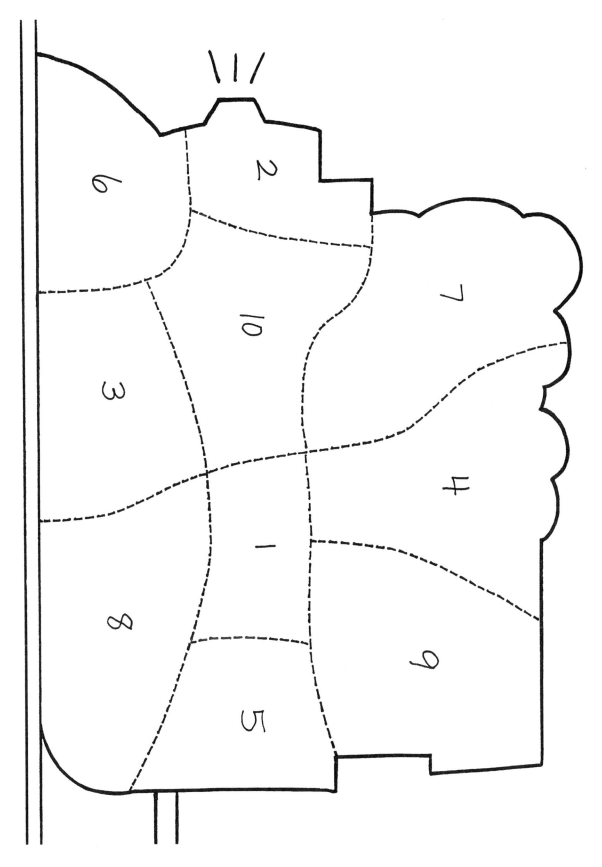

Addition and Subtraction 0-10

Name _____

Addition and Subtraction 0–10

Name _____

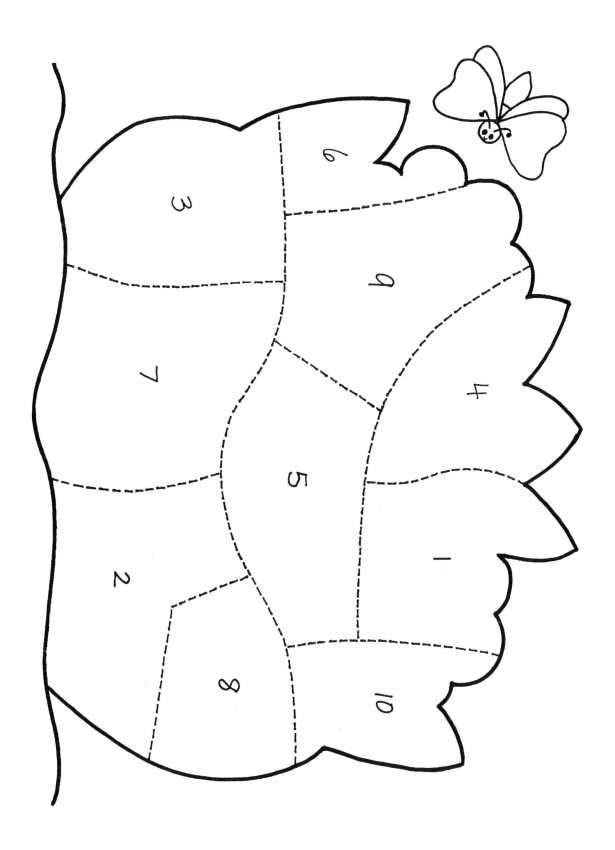

Addition and Subtraction 0-10

Answer Key

Puppy p. 6

Dinosaur p. 8
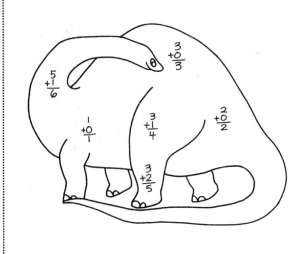

Bear p. 10
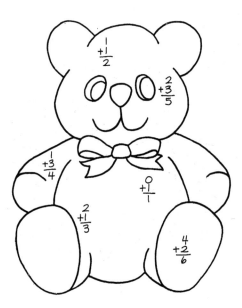

Whale p. 12
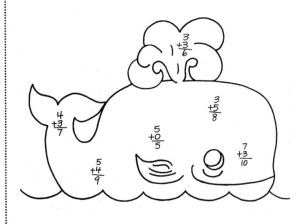

Answer Key

Elephant p. 14

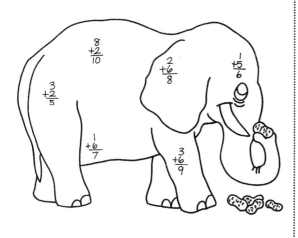

Snowman p. 16

Turtle p. 18

Frog p. 20

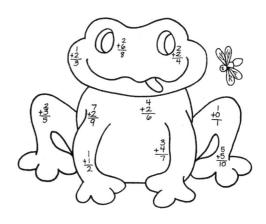

Answer Key

Tugboat p. 22

Jack-'o-lantern p. 24

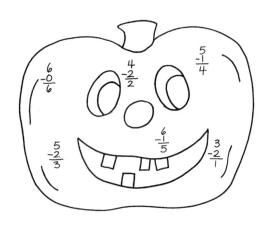

Rocket p. 26

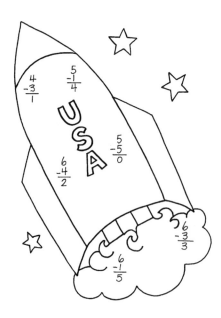

Turkey p. 28

Answer Key

Fish p. 30

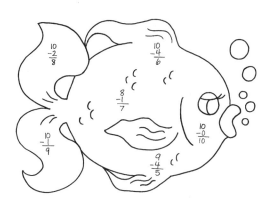

Humpty Dumpty p. 32

Duck p. 34

Bee p. 36

Answer Key

School Bus p. 38

Train p. 40

Flowers p. 42

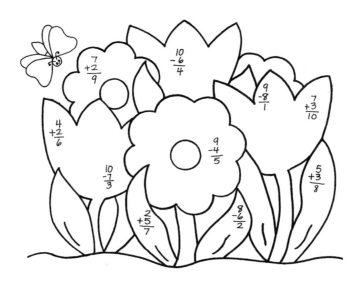